Decimals, Fractions & Percentages

BGCSE CORE LEVEL MATHEMATICS GUIDE

Decimals, Fractions and Percentages: BGCSE Core Level Mathematics Guide by Zemi Stewart, 2018

This book or parts thereof may not be reproduced in any form, stored in a retrieval system, or transmitted in any form by any means—electronic, mechanical, photocopy, recording, or otherwise—without prior written permission of the author.

All inquiries regarding this publication, ordering information, requests for usage rights or corrections should be sent by email to mathwithmisszemi@gmail.com.

While the author has made every effort to provide accurate Internet addresses at the time of publication, neither the publisher nor the author assumes responsibility for errors or changes that occur after publication.

ISBN-13: 978 1725 1854 4
ISBN-10: 1725518546

Printed and bound in the United States of America by CreateSpace

Table of Contents

Welcome	5
About the BGCSE	6
Core Syllabus	7
Core Expectations	8
Introduction	8
Fractions	**9**
Exercise 1	9
Adding and Subtracting Fractions	10
Exercise 2	10
Multiplying and Dividing Fractions	12
Exercise 3	13
Fractions of Quantity	13
Exercise 4	14
Equivalent Fractions	15
Exercise 5	15
Comparing Fractions	16
Decimals	**16**
Adding and Subtracting Decimals	17
Exercise 6	17
Multiplying Decimals	17
Exercise 7	18
Dividing Decimals	18
Exercise 8	19
Exercise 9	20

Converting Decimals to Fractions	20
Converting Fractions to Decimals	20
Exercise 10	21

Percentages — 22

Exercise 11	22
Expressing a Quantity as a Percentage of Another	23
Calculating the Percentage of a Quantity	23
Exercise 12	24
Value-added Tax	24
Exercise 13	25
Percentage Increase	25
Percentage Decrease	25
Exercise 14	26

Exam Style Questions — 27

Resources — 35

Welcome

Hi Student or Parent,

Welcome to this short guide on Decimals, Fractions & Percentages. The material in this guide covers the core level of the Bahamas General Certificate of Secondary Education (BGCSE) Mathematics examination. I have tried to present the material in a way that is easy to understand and fun to read. Remember, Math can be studied, not just practiced. I hope you enjoy this guide and many others in this series.

-Zemi Stewart
Founder, Math with Miss Zemi

About the BGCSE

The Bahamas General Certificate of Secondary Education (BGCSE) was formed by the Bahamas' Ministry of Education and commissioned by Cambridge University in the United Kingdom. BGCSE examinations are based on the United Kingdom General Certificate of Secondary Education (GCSE) and the International General Certificate of Secondary Education (IGCSE). The Mathematics examination is split into two levels: Core and Extended.

Core Syllabus

- ✓ Number
- ✓ Set Notation & Language
- ✓ Square, Square Root, Cube & Cube Root
- ✓ Directed Numbers
- ✓ Decimals, Fractions and Percentages
- ✓ Ordering (Comparing Quantities)
- ✓ Scientific Notation
- ✓ The Four Rules
- ✓ Estimation & Limits of Accuracy
- ✓ Ratio, Proportion & Rate
- ✓ Use of Calculator
- ✓ Measures
- ✓ Time
- ✓ Money
- ✓ Personal & Household Finance
- ✓ Graphs in Practical Situations
- ✓ Graphs of Functions
- ✓ Algebraic Representation & Formulae
- ✓ Algebraic Manipulation
- ✓ Indices (Powers/Exponents)
- ✓ Algebraic Equations
- ✓ Linear Inequalities and Regions
- ✓ Symmetry
- ✓ Geometrical Terms & Relationships
- ✓ Geometrical Construction
- ✓ Angle Properties
- ✓ Measurement (Area, Volume, etc.)
- ✓ Trigonometry
- ✓ Statistics
- ✓ Probability
- ✓ Vectors in Two Dimensions
- ✓ Transformations

> Yes, that is a whole lot of material, but do not think for a second that you cannot learn and master all of it. Never limit yourself!

Core Expectations

The examiners expect the student to be able to:
- Use the language and notation of simple vulgar (common) and decimal fractions and percentages in appropriate contexts.
- Recognize equivalence between these forms and convert from one to the other.
- Calculate a given percentage of a quantity.
- Express one quantity as a percentage of the other.
- Calculate percentage increase or decrease.

Introduction

Decimals, fractions and percentages are used throughout our lives. We encounter them in many real-world scenarios such as, but not limited to, managing personal finances, during shopping trips and in cooking & baking.

Fundamentally, decimals, fractions and percentages are simply different ways of expressing or showing the same value. This guide will build on your foundational knowledge of fractions, decimals and percentages and extend your knowledge to the BGCSE Core Level.

Fractions

A fraction shows parts of a whole and is expressed as $\frac{a}{b}$, where "a" and "b" are two whole numbers.

$$\frac{2}{4} \longrightarrow \textbf{numerator}$$
$$\phantom{\frac{2}{4}} \longrightarrow \textbf{denominator}$$

The numerator represents how many parts there are while the denominator represents how many equal parts the whole is divided into. The fraction above shows 2 parts out of 4. Note, $\frac{2}{4}$ is the same as 2 ÷ 4.

Exercise 1

A popular nursery rhyme reads:

> "30 days hath September, April, June and November. All the rest have 31, except for February alone. It has 28 days clear, and 29 in each Leap Year."

You would be surprised how often I have used this nursey rhyme to solve math problems. If you have trouble remembering how many days are in each month, I suggest you remember it also.

(1) Memorize the nursey rhyme above.

(2) In April there were 14 rainy days. What fraction of April was rainy?

There are 30 days in April so the denominator is 30. If you guessed $\frac{14}{30}$ you are correct!

(2) Your little brother just received his BJC results. He obtained 3 As, 2 Bs and 1 C. What fraction of his exam results were As?

There are several types of fractions that you should know:
1) proper fractions, e.g. $\frac{1}{2}$
2) improper fractions, e.g. $\frac{11}{3}$
3) mixed fractions, e.g. $3\frac{1}{8}$

Adding and Subtracting Fractions

There are a few key rules you should remember when adding and subtracting fractions:

- If the denominators are the same, add or subtract the numerators and simplify.
- If the denominators are not the same, find the lowest common multiple (LCM) of the denominators (also called the lowest common denominator or LCD) and simplify.
- When dealing with mixed fractions, it is normally easier to change them to improper fractions first; find the LCD and then add or subtract the numerators.

I will provide a few examples and exercises below; however, for more details on this subject, please review the materials listed in the Resources section of this guide.

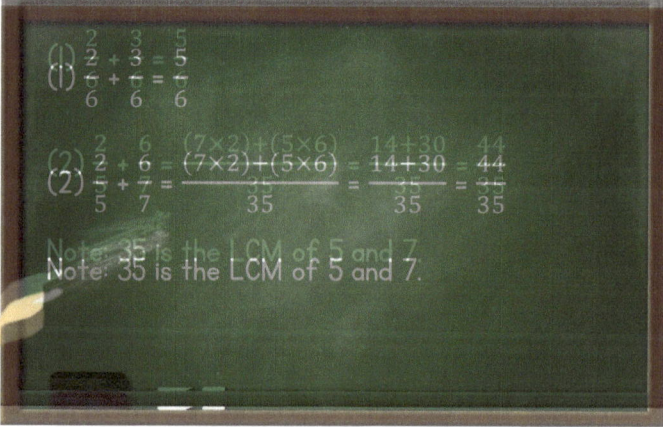

(1) $\frac{2}{6} + \frac{3}{6} = \frac{5}{6}$

(2) $\frac{2}{5} + \frac{6}{7} = \frac{(7 \times 2)+(5 \times 6)}{35} = \frac{14+30}{35} = \frac{44}{35}$

Note: 35 is the LCM of 5 and 7.

Exercise 2

(1) Evaluate the following:

(a) $\frac{1}{2} - \frac{1}{4}$

(b) $1\frac{5}{6} + 3\frac{3}{2}$

(c) $\frac{2}{4} = \frac{4}{5}$ (d) $\frac{2}{5} - \frac{3}{5}$

(2) Jordan is involved in many extracurricular activities. He spends $\frac{1}{6}$ of his time playing baseball, $\frac{1}{3}$ of his time taking piano lessons, and $\frac{1}{9}$ of his time at swimming practice. The remainder of his time is spent doing other activities.

Baseball	Piano Lessons	Swimming Lessons	Other Activities
$\frac{1}{6}$	$\frac{1}{3}$	$\frac{1}{9}$	

Calculate:

(a) The total fraction of his time occupied.

(b) The total fraction of his time doing other activities.

(3) Work out $\frac{5}{12} - \frac{1}{4}$. Write your answer in simplest form.

Multiplying and Dividing Fractions

Below are the key rules for multiplying and dividing fractions:

- When multiplying <u>improper</u> or <u>proper</u> fractions, multiply the numerators (top numbers) and multiply the denominators (bottom numbers), then simplify.
- When multiplying <u>mixed numbers</u>, convert to an improper fraction then multiply the numerators, and denominators.
- When dividing, replace the division sign with a multiplication sign and multiply the first fraction by the reciprocal of the second.

To obtain the "reciprocal" of a fraction simply flip it. For example, $^1/_2$ becomes $^2/_1$.

Examples:

(1) $\frac{3}{5} \times 3\frac{3}{5} = \frac{3}{5} \times \frac{18}{5} = \frac{54}{25}$

(2) $\frac{3}{5} \div 2 = \frac{3}{5} \div \frac{2}{1} = \frac{3}{5} \times \frac{1}{2} = \frac{(3 \times 1)}{(5 \times 2)} = \frac{3}{10}$

The fraction in (1) cannot be simplified further as 54 and 25 have no common factors (or divisors). Likewise, (2) is in its simplest form.

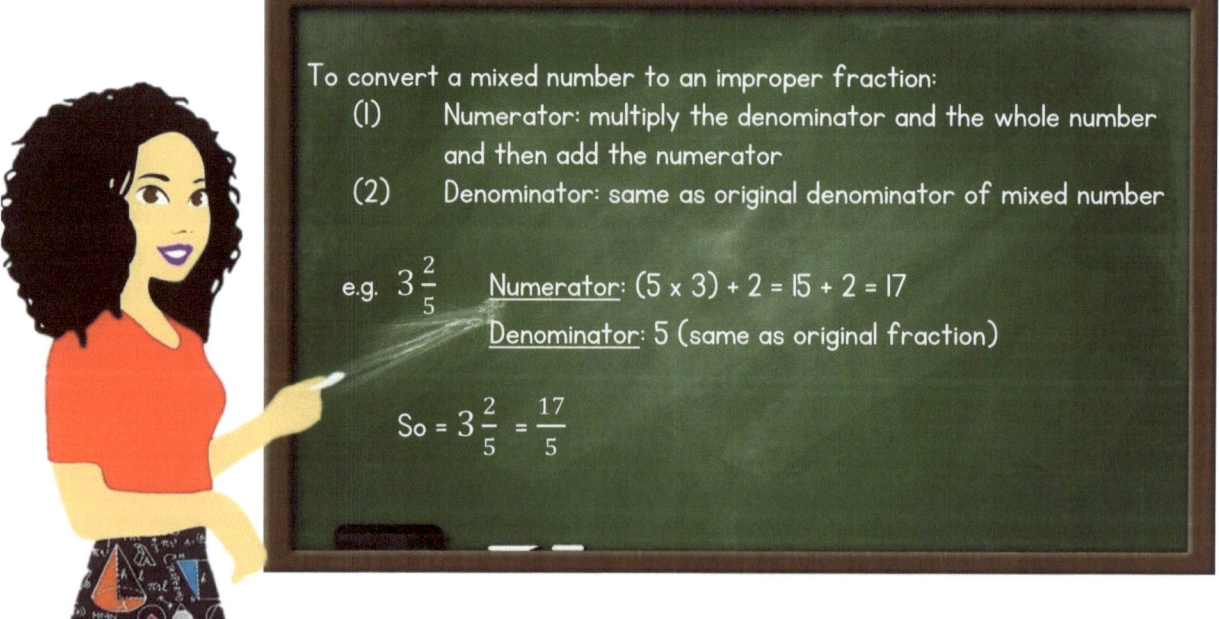

Let's put into practice what you've learned. Try the exercises on the following page. If any of the exercises seem difficult for you, review this section again or review the materials in the Resources section.

Exercise 3

(1) Evaluate the following:

(a) $\frac{5}{6} \times \frac{1}{4}$

(b) $\frac{1}{7} \times \frac{3}{5}$

(c) $4\frac{4}{5} \times 1\frac{5}{8}$

(d) $4 \times 3\frac{3}{8}$

(2) Evaluate the following and give your answer as a mixed fraction:

(a) $1\frac{1}{2} \times \frac{21}{2}$

(b) $2\frac{1}{2} \times 3\frac{1}{4}$

Fractions of Quantity

There are multiple methods to determine fractions of quantities (also called "fractions of amounts"); however, this guide will focus on the most commonly used method: multiplying fractions.

Q. How do we find $3/5$ of 20?

This is the same as $3/5 \times 20$.

$3/5 \times 20 = 3/5 \times 20/1 = \frac{(3 \times 20)}{(5 \times 1)} = 60/5 = 12$

In "fractions of quantity" questions, replace the "of" with a multiplication sign (as demonstrated in the example above).

Your turn!
Q. What is $3/7$ of 35?

Exercise 4

(1) $\frac{3}{4}$ of a pan of macaroni is sitting on the counter. You decide to eat $\frac{1}{3}$ of it with your Sunday dinner. How much of the whole pan did you eat?

(2) You have 24 pairs of jeans and want to donate half of them to the Salvation Army. How many do you donate?

(3) Find three quarters of $1.
Tip: Convert $1 to cents first.

(4) Which is larger, $3/5$ of 24 or $3/4$ of 20?
Show your working.

Always be careful of the units! For example, 20 cents as a fraction of $2 is not $\frac{20}{2}$ but rather $\frac{20}{200}$ (because $1.00 = 100$ cents and so $2 = 200$ cents). We converted the $2 to cents so that the units of the two quantities would be the same.

Q: What is 30 cm as a fraction of 5m?
A: 30cm as a fraction of 5m is $\frac{30}{500}$ (since 5m = 500 cm).

Equivalent Fractions

A fraction can be written in different ways and still mean the same thing. These are called equivalent fractions.

As you can see below, each shaded region represents one half of a circle. Therefore, $\frac{1}{2}, \frac{2}{4}, \frac{3}{6}$ and $\frac{4}{8}$ are all equivalent fractions.

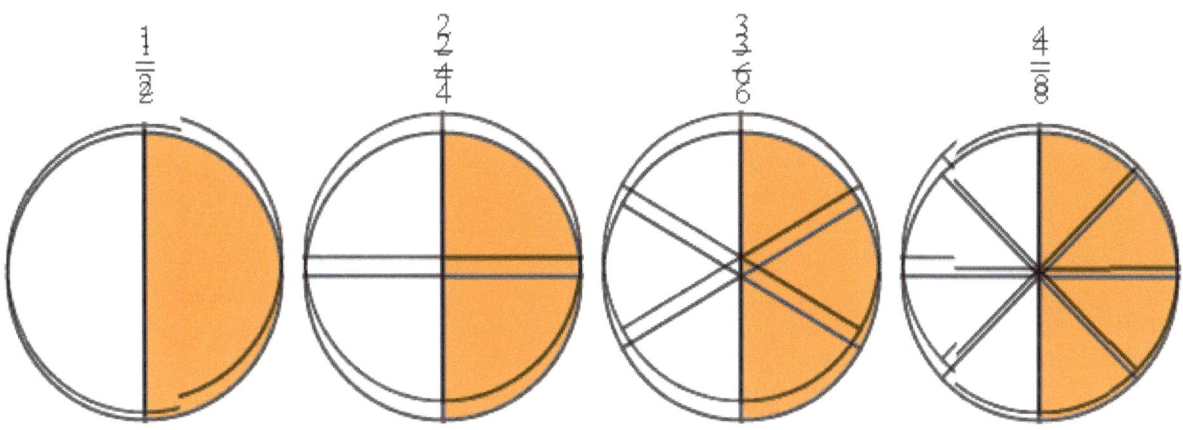

You can produce an infinite number of equivalent fractions. How? Simply multiply the numerator and the denominator by the same number.

For example, to find fractions that are equivalent to $\frac{1}{5}$ multiply the numerator and denominator by the same number.

$\frac{1}{5} = \frac{2}{10}$ (multiply numerator and denominator by 2)

$\frac{1}{5} = \frac{3}{15}$ (multiply numerator and denominator by 3)

You can go on and on because a single fraction has an infinite number of equivalent fractions.

Exercise 5

(1) Ian and Zemi ordered pizzas for dinner. Ian ate $\frac{1}{3}$ of his pizza while Zemi ate $\frac{3}{9}$ of hers. Did they eat an equivalent amount of pizza?

(2) Lauren ate $\frac{1}{2}$ of her Snickers bar. Javon ate $\frac{1}{5}$ of his Mars bar. Did they eat an equal amount of their chocolate bars?

Comparing Fractions

When comparing the size of fractions, you must find equivalent fractions that have the same denominator.

Q. Which is bigger, $\frac{5}{8}$ or $\frac{4}{7}$?

Their denominator must be a number that both 8 and 7 can divide into. I chose the lowest common multiple (LCM) of 8 and 7 which is 56. In doing so, I note:

$$\frac{5}{8} = \frac{35}{56} \text{ and } \frac{4}{7} = \frac{32}{56}$$

We can immediately see which fraction is bigger (and smaller) because the fractions share the same denominator. But be careful! Your final answer should be "$\frac{5}{8}$ is bigger", *not* "$\frac{35}{56}$ is bigger". They are equivalent fractions so the answer is technically true, but always answer exactly what you are asked.

Now you try. Arrange the following fractions in ascending order: $\frac{7}{10}, \frac{1}{2}, \frac{14}{25}, \frac{3}{5}$.

Note: "Ascending order" means to order the numbers from smallest to greatest. "Descending order" means to order the numbers from largest to smallest.

Decimals

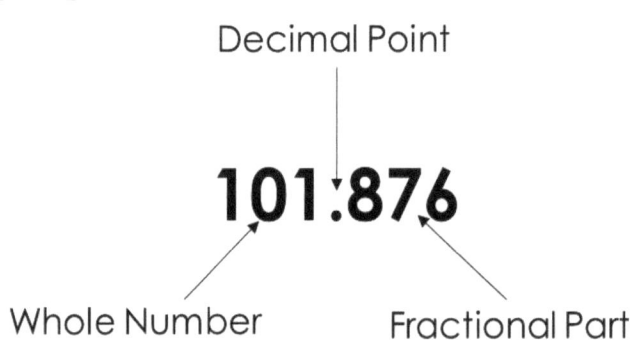

As in the example above, we use a decimal point to separate the whole number part from the fractional part. The fraction part is also known as "parts of a whole" (tenths, hundredths, thousandths, etc.).

- A **tenth** is 1/10 of a unit (0.1 in decimal form)
- A **hundredth** is 1/100 of a unit (0.01 in decimal form)
- A **thousandth** is 1/1000 of a unit (0.001 in decimal form)

Adding and Subtracting Decimals

When adding and subtracting decimals, add or subtract as normal, but make sure that you keep decimal points aligned.

Exercise 6

Try the simple exercises below to see if you remember the basics of decimal computation.

1) 4.27 + 2.30

2) 5 – 0.24

Multiplying Decimals

When multiplying decimals there are two key rules that you must keep in mind:
- (1) Multiply as normal (pretend the decimal point does not exist).
- (2) When you have obtained the answer, place the decimal point in the correct position.

Q. *What is the correct position?*

The product (result of the multiplication) should have as many decimal places as the two numbers you were multiplying combined. Let's illustrate this with an example to be sure you've got it.

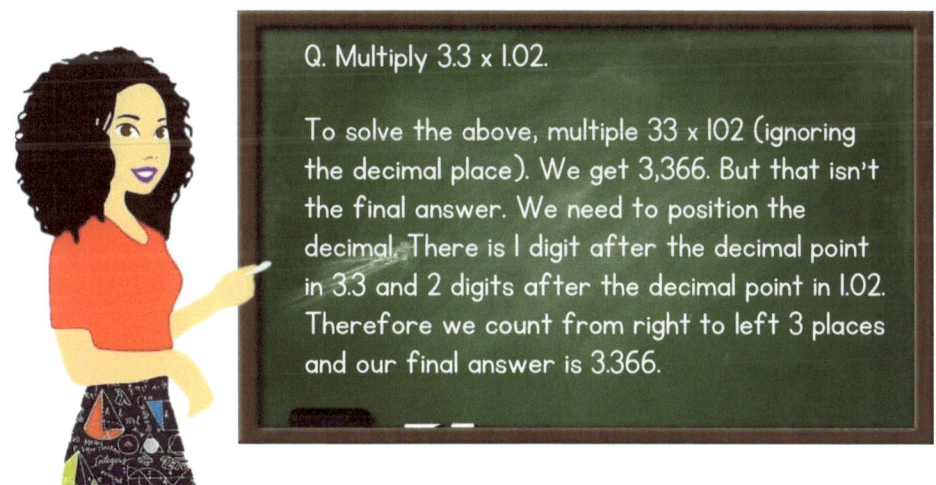

Q. Multiply 3.3 x 1.02.

To solve the above, multiple 33 x 102 (ignoring the decimal place). We get 3,366. But that isn't the final answer. We need to position the decimal. There is 1 digit after the decimal point in 3.3 and 2 digits after the decimal point in 1.02. Therefore we count from right to left 3 places and our final answer is 3.366.

Tip: When multiplying by multiples of 10, move the decimal place to the right n times, where n is the number of 0s in the number. When dividing by multiples of 10, move the decimal place to the left n times, where n is the number of 0s in the number.

Example:
$6.23 \times 1000 \rightarrow$ move the decimal place to the right 3 times (because 1,000 has 3 zeros)

Therefore, $6.23 \times 1000 = 6230.0$ or $6,230$.

→ Move the decimal 3 spaces to the right to get 6230.0 or 6,230.

Exercise 7

(1) What is 1.2×1.1?

(2) Solve 3.42×2.

(3) What is 3.45×2.1?

Dividing Decimals

When dividing a decimal by a whole number, divide as usual but keep the decimal points aligned.

Example:

$3.75 \div 3 = 3\overline{)3.75}^{1.25} = 1.25$ (correct to 2 decimal places)

Exercise 8

(1) Calculate and give your answers correct to 2 decimal places:

a. $0.95 \div 21$

b. $9.1 \div 7$

c. $12.4 \div 6$

When dividing a decimal by another decimal, you need to use equivalent fractions.

Example:

$2.42 \div 0.2$

$2.42 \div 0.2$ is the same as $24.2 \div 2$

How? We have multiplied both the numerator and denominator by 10.

$2.42 \times 10 = 24.2$
$0.2 \times 10 = 2$

> Remember the trick we learned about multiplying by powers of 10? You can apply it here.

Again, $3.715 \div 0.005$ is the same as $3715 \div 5$. In this case, we multiplied both sides by 1000.

Tip: Always multiply the numerator and denominator by the same number, and make sure that the denominator is a whole number.

Exercise 9

(1) Evaluate correct to 3 significant figures.

$$\frac{0.0785}{0.249}$$

Note: Check the resources section for additional information on significant figures. "Correct to 3 significant figures" is not to be confused with the phrase "correct to 3 decimal places".

(2) Evaluate:
   a. $0.9 \div 0.009$

   b. $1.35 \div 0.15$

   c. $0.04 \div 0.2$

Converting Decimals to Fractions (and vice versa)

Converting fractions to decimals is often easier to compute, so we'll start there.

$$\frac{5}{3} = 5 \div 3 = 1.6666....$$

Note: This is a recurring decimal because the "6" repeats. Another way of writing this would be $1.\bar{6}$. We place the line over the repeating digit. If we were to round to 3 decimal places, the answer would be 1.667.

Q. Convert $\frac{4}{25}$ to a decimal.

$$\frac{4}{25} = 4 \div 25 = 0.16$$

When converting decimals to fractions, we must rely on our foundational knowledge of decimals.

0.2 means "two tenths" or $\frac{2}{10}$

0.23 means "23 hundredths" or $\frac{23}{100}$

0.543 means "543 thousandths" or $\frac{543}{1000}$

...and so on.

Don't worry about converting recurring decimals to fractions. This is not part of the Core curriculum and I have yet to see it as part of the Extended curriculum.

Q. What about 2.356? What is this as a fraction?

$2.356 = 2\frac{356}{1000} = 2\frac{89}{250}$ (remember to give your answer in simplest form)

Exercise 10

(1) Which is bigger, $1\frac{4}{5}$ or 1.75? Show your working.

(2) Convert 3.53 to a fraction.

(3) Convert 0.657 to a fraction.

(4) Memorize these conversions.

| ¼ = 0.25 | ½ = 0.5 | ¾ = 0.75 |

Percentages

A percentage is a number or ratio expressed as a fraction of 100. That is, the denominator of a percentage is always 100. Percentages are often denoted using the percent sign, %.

65% means 65 out of 100

Therefore, as a fraction, $65\% \equiv \frac{65}{100} \equiv \frac{13}{20}$

and as a decimal, $65\% \equiv \frac{65}{100} \equiv 0.65$

Similarly, the fraction $\frac{7}{20}$ as a percentage equals $\frac{7}{20} \times 100 \equiv 35\%$

and $\frac{7}{20} \equiv \frac{35}{100} \equiv 0.35$.

Exercise 11

Complete the following table:

Fraction	Decimal	Percentage
$\frac{1}{4}$		
	0.1	
		50%
$\frac{3}{4}$		
		20%
	0.3	
$\frac{9}{10}$		
		60%

Please show your working:

Expressing a Quantity as a Percentage of Another

Let's begin by reviewing the example below:

Q: Express 45 cents as a fraction of $1.35.
1.35 in cents = 135
Therefore, the fraction would be $\frac{45}{135}$.
Let's express this answer in lowest terms.
$\frac{45}{135} = \frac{9}{27} = \frac{3}{9} = \frac{1}{3}$
45 cents as a fraction of $1.35 is therefore $\frac{1}{3}$.

Be careful with the units when solving such questions. Notice, I changed the dollar value to cents before answering the question. A similar approach is taken in the example below:

Q: Express 1200 g as a percentage of 3 kg.
Grams and kilograms are clearly different units. We therefore have to convert both of these quantities to the same unit before answering the question.
1 kg = 1000 g
Therefore, 3 kg = 3000 g
The fraction is therfore, $\frac{1200}{3000} = \frac{12}{30} = \frac{2}{5}$.

Calculating the Percentage of a Quantity

Finding the percentage of a quantity is very practical for day-to-day life. Sales such as "20% off" or "50% off", and even Value-Added Tax (VAT) calculations are all real-life applications of calculating the percentage of a quantity.

Q: Find 40% of $50.
This statement is equivalent to $\frac{40}{100} \times \$50$.
$\frac{40}{100} \times \$50 = \$20$

Q: If 20% of the children in a class play the violin, what percentage do not?
The percentage who play is 20%, therefore the percentage who do not is 100% - 20% = 80%.

Exercise 12

(1) Jami is buying a pair of jeans. The original price was $75, but there is a discount of 30%. How much will the discount be?

(2) You are on a plane to Singapore. Seventy-six percent of the passengers on the flight are Malaysian. What percentage of them are not?

(3) Deductions from a man's wage were: health insurance 21%, pension 8%, and other deductions 7%. What percentage did he keep?

(4) At St. James High, 35% of the students take woodwork, 25% take technical drawing and 20% take both subjects. What percentage of the students study neither?

(5) Find 24% of 7.5 m

(6) Find $12\frac{1}{2}$% of 4.88 cm

Value Added Tax (VAT)

Value Added Tax in the Bahamas is charged at a rate of 12% (as of 2018). Previously, it was calculated at a rate of 7.5%.

Exercise 13

Javon wants to purchase a car that costs $4,500 before VAT.
(a) Work out the cost of the VAT that is charged at 12%.

(b) In the past, the VAT would have been calculated at 7.5%. How much more does Javon pay in VAT at the current rate?

Percentage Increase

If a quantity or amount increases by a certain percentage, we multiply the quantity or amount by the multiplying factor to determine the increase.

Example:
A pair of Fashion Nova jeans cost $23 dollars on sale, but will increase by 20% tomorrow (when the sale is over).

The increased amount is (100 + 20)% or 120% of the original cost (which in this case is the sale price). 120% is the multiplying factor.

We calculate the increase as follows:

$\frac{120}{100}$ x $23 = $27.60

More generally, for "percentage increase", the multiplying factor is
(100 + percentage increase)%.

Percentage Decrease

Similarly, for "percentage decrease" calculations, the multiplying factor is
(100 − percentage decrease)%.

Exercise 14

(1) Decrease 100 by 30%.

(2) The BPL bill increased by 20% this month. Last month the bill was $175.55. What is the bill now?

(3) Last year 250 high school students failed the Office Procedures exam. Examiners reported that the number of fails fell by 16%. How many students failed this year?

Exam Style Questions

1. Express 48% as a
   a. fraction in its lowest terms [2]

   b. decimal [1]

2. a. Use your calculator to work out the exact value of
   $$\frac{32.41 \times 1.382}{8.4}$$
   Tip: Input ((32.41 × 1.382) ÷ 8.4) into your calculator. [1]

   b. Express the answer in (a) correct to:
      (i) two decimal places [1]

      (ii) three significant figures [1]

3. David and Deandre are close friends who attend the same school. David's home is 8.87 miles away from the school. Deandre's home is 3 times as far as David's home from school. What is the distance between Deandre's school and his home?

4. Your mother sent you to the store for tomato paste. A 6 oz can of tomato paste usually costs $3.26; however, you see an 8 oz can on special for $3.01.
   a. How much do you save by purchasing the 8 oz can?

   b. What would your change be from $5.00?

5. A box contains 560 g of cereal. A family pack of cereal contains an extra 35% more at the same price. How many grams of cereal are in the family pack?

[3]

6. Write 12.24 correct to
    a. the nearest whole number _____ [1]

    b. one decimal place _____ [1]

    c. the nearest ten _____ [1]

    d. the nearest tenth _____ [1]

7. House of Hall, an online boutique, sold a dress for $189.50. This was $\frac{1}{5}$ more than the cost price of the dress. What is the cost price?

[2]

8. Write 0.6 as:
    a. a fraction [1]

    b. a percentage [1]

9. Gabrielle earns $26,000 per year. She gets an increase of 3%.
    a. How much money is the increase per year? [2]

    b. How much money is the increase per month? [1]

10. Using the given fractions, $5\frac{1}{6}$ and $\frac{1}{2}$, [4]
    a. (i) change the mixed number fraction to an improper fraction

    (ii) state the Lowest Common Denominator (LCD) of the fractions.

    b. Use the LCD to find the sum of the fractions.

    c. Write your answer to (b) as a fraction in its lowest terms.

11. A calculator display shows 4768.29.
    Write this number correct to
    a. the nearest hundred [1]

    b. one decimal place [1]

    c. one significant figure [1]

12. Express as a fraction in lowest terms,
    a. 8% [2]

    b. 0.475 [2]

13. Express $\frac{1}{5}$ as a decimal. [1]

14. Convert
    a. 75% to a fraction in lowest terms [1]

    b. 0.075 as a percentage [1]

    c. $7\frac{1}{8}$ to a decimal [1]

15. Write down the value of $(12.52)^2$
    a. exactly [1]

    b. to two significant figures [1]

    c. to one decimal place [1]

    d. to the nearest hundred [1]

16. Simplify $\frac{15+7}{13-2}$ [1]

17. Calculate the fraction that is halfway between $\frac{1}{4}$ and $\frac{1}{2}$. [1]
Tip: You may consider using a number line to solve this. An alternative is to use your calculator.

18. Express as a fraction in lowest terms;
    a. $\frac{35}{120}$ [2]

    b. 6% [1]

c. 0.375 [2]

19. Write down the value of $(11.36)^2$
   a. exactly [1]

   b. to two decimal places [1]

20. a. Express the decimal number 0.375 as
   (i) a percentage [1]

   (ii) a fraction in lowest terms [2]

   b. A class project requires 24 pieces of copper wire, each a length of $1\frac{3}{5}$ inches. Assuming there is no waste in cutting, calculate the minimum length of the wire that should be purchased. [2]

21. Arrange the following decimals in order from smallest to largest: [2]
    0.02; 0.0042; 0.0301; 0.036

22. Jackie obtained an answer of 0.087615 on her calculator. She rounded the number off to 0.0876. State the amount of:
    (i) decimal places she rounded to [1]

    (ii) significant figures she rounded to [1]

23. Express the fraction $\frac{21}{63}$
    a. in lowest terms [1]

    b. in decimal form [1]

    c. as a percent [2]

24. Insert one of the symbols <, >, = below to make each statement true.

    a. $\dfrac{2}{3}$ ____ 0.6 [1]

    b. 8% ____ 0.8 [1]

    c. $\dfrac{3}{8}$ ____ $\dfrac{3}{12}$ [1]

    d. $\dfrac{1}{4}$ ____ 20% [1]

25. For the fractions $\dfrac{1}{2}$, $\dfrac{3}{4}$, $\dfrac{1}{3}$, and $\dfrac{2}{5}$

    a. Write the Lowest Common Denominator [1]

    b. List the fraction in ascending order. (Show your working.) [3]

26. Calculate the difference in length between $2\dfrac{1}{4}$ and $1\dfrac{7}{8}$. [2]

27. Complete the table. [3]

Fraction	Decimal	Percentage
½		50%
	0.75	
	0.10	10%

28. Paul buys 30 cans of soda for a party. [4]
    He is given a 20% discount off the total price.

    One can costs 75 cents **before** the discount.

    How much does he pay?

29. Work out 0.6×0.1 [1]

30. Work out $0.5 - 0.18$ [2]

31. Work out $\frac{5}{6} \times \frac{3}{20}$. Express your answer as a fraction in simplest form.   [3]

32. Jami needs $\frac{2}{3}$ of a tank of gas to drive home. She has $\frac{5}{8}$ of a tank. Does she have enough gas to get home or should she stop at the gas station along the way?   [2]

Show your working. Give your final answer in complete sentences.

33. Evaluate   [2]

$$\frac{3}{5} \times \frac{2}{4} \div \frac{3}{16}$$

34. I completed $\frac{2}{3}$ of my project last month and $\frac{1}{4}$ of my project this week.
  a. What fraction of my project is complete?   [3]

  b. What fraction of my project remains to be completed?   [1]

35. a. Anwar scored 17 out of 20 on his Chemistry test. Calculate this as a percentage. [2]

b. On another test, he scored 75%. If that test was out of 24 marks, how many marks did he receive? [2]

Resources

Fractions
https://www.bbc.com/education/guides/z2hsrwx/revision/1
http://studyjams.scholastic.com/studyjams/jams/math/index.htm
https://revisionmaths.com/gcse-maths-revision/number/fractions

Decimals
https://revisionmaths.com/gcse-maths-revision/number/decimal-numbers
http://studyjams.scholastic.com/studyjams/jams/math/index.htm
http://www.bbc.co.uk/schools/gcsebitesize/maths/number/decimalsrev1.shtml

Percentages
https://www.mathsisfun.com/percentage.html
http://www.bbc.co.uk/schools/gcsebitesize/maths/number/fracsdecpersrev1.shtml
https://mathsmadeeasy.co.uk/gcse-maths-revision/percentages-gcse-revision-and-worksheets/

Significant Figures
http://www.bbc.co.uk/schools/gcsebitesize/maths/number/roundestimaterev3.shtml

www.ingramcontent.com/pod-product-compliance
Lightning Source LLC
Chambersburg PA
CBHW051935210526
45473CB00006B/2262